YOUR KNOWLEDGE HAS VALUE

AF152195

- We will publish your bachelor's and
  master's thesis, essays and papers

- Your own eBook and book -
  sold worldwide in all relevant shops

- Earn money with each sale

Upload your text at www.GRIN.com
and publish for free

GRIN

John Bredakis

The mighty role of right angle triangles to integral calculus

GRIN Publishing

Bibliographic information published by the German National Library:

The German National Library lists this publication in the National Bibliography;
detailed bibliographic data are available on the Internet at http://dnb.dnb.de .

This book is copyright material and must not be copied, reproduced, transferred,
distributed, leased, licensed or publicly performed or used in any way except as
specifically permitted in writing by the publishers, as allowed under the terms and
conditions under which it was purchased or as strictly permitted by applicable
copyright law. Any unauthorized distribution or use of this text may be a direct
infringement of the author s and publisher s rights and those responsible may be
liable in law accordingly.

Imprint:

Copyright © 2011 GRIN Verlag, Open Publishing GmbH
Print and binding: Books on Demand GmbH, Norderstedt Germany
ISBN: 978-3-640-89631-8

This book at GRIN:

http://www.grin.com/en/e-book/170468/the-mighty-role-of-right-angle-triangles-
to-integral-calculus

GRIN - Your knowledge has value

Since its foundation in 1998, GRIN has specialized in publishing academic texts by students, college teachers and other academics as e-book and printed book. The website www.grin.com is an ideal platform for presenting term papers, final papers, scientific essays, dissertations and specialist books.

Visit us on the internet:

http://www.grin.com/

http://www.facebook.com/grincom

http://www.twitter.com/grin_com

John Bredakis

The mighty role of right angle triangles to integral calculus

Essay

G R I N

Verlag fur akademische Texts

The mighty role of right angle triangles to integral calculus

Integrals with integrand functions fitting into the perimeter of those triangles are solvable by trigonometric substitution, like the exponential and hyperbolic integrals.

Introduction:
Bypassing the inverse functions by right angle triangles:

u = arc sin (x/a)	u = arc csc (x/a)	u = arc tan (x/a)

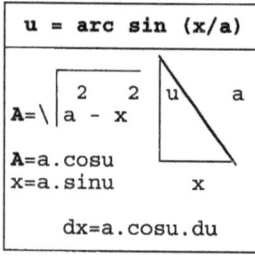

$A=\sqrt{a^2 - x^2}$

A=a.cosu
x=a.sinu

dx=a.cosu.du

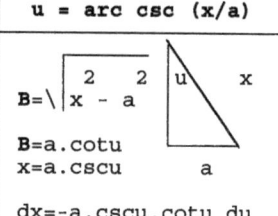

$B=\sqrt{x^2 - a^2}$

B=a.cotu
x=a.cscu

dx=-a.cscu.cotu.du

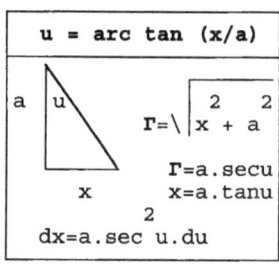

$\Gamma=\sqrt{x^2 + a^2}$

Γ=a.secu
x=a.tanu

dx=a.sec²u.du

d/dx arc sin (x/a)	d/dx arc csc (x/a)	d/dx arc tan (x/a)

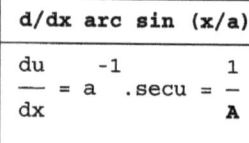

$\dfrac{du}{dx} = a^{-1}.secu = \dfrac{1}{A}$

$\dfrac{du}{dx}=-a^{-1}.sinu.tanu=\dfrac{-a}{x.B}$

$\dfrac{du}{dx} = a^{-1}.cos^2u = \dfrac{a}{\Gamma^2}$

- - - - - - - - - - -

v = arc cos (x/a)	v = arc sec (x/a)	v = arc cot (x/a)

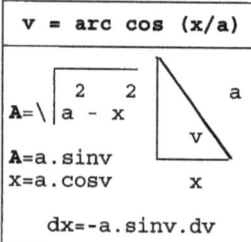

$A=\sqrt{a^2 - x^2}$

A=a.sinv
x=a.cosv

dx=-a.sinv.dv

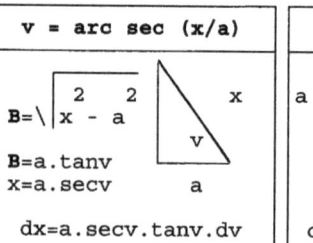

$B=\sqrt{x^2 - a^2}$

B=a.tanv
x=a.secv

dx=a.secv.tanv.dv

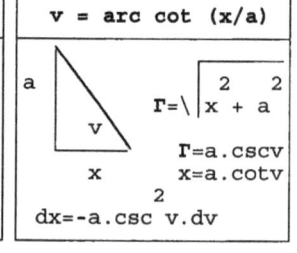

$\Gamma=\sqrt{x^2 + a^2}$

Γ=a.cscv
x=a.cotv

dx=-a.csc²v.dv

d/dx arc cos (x/a)	d/dx arc sec (x/a)	d/dx arc cot (x/a)

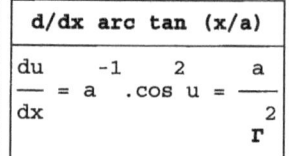

$\dfrac{dv}{dx} = -a^{-1}.cscv = \dfrac{-1}{A}$

$\dfrac{dv}{dx} =a^{-1}.cosv.cotv=\dfrac{a}{x.B}$

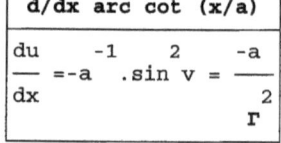

$\dfrac{du}{dx} =-a^{-1}.sin^2v = \dfrac{-a}{\Gamma^2}$

J.K.Bredakis MD

Properties of transcendental functions

<table>
<tr>
<td>

A. Properties of exponential and logarithmic functions

</td>
<td>

From $\log(a)x = y$ we get that $a^y = x$
base a

</td>
</tr>
<tr>
<td>

$e^a \cdot e^b = e^{(a+b)}$

</td>
<td>

$\log_{\text{base a}} x = \dfrac{\ln x}{\ln a}$ $\log_{10} x = \dfrac{\ln x}{\ln 10} = 2.306$

</td>
</tr>
<tr>
<td>

$e^a / e^b = e^{(a-b)}$

</td>
<td>

$\ln 1 = 0$ $\ln e = 1$
$\log e = 1/2.306 = 0.432429$

</td>
</tr>
<tr>
<td>

$\left[e^a \right]^b = e^{a.b}$

</td>
<td>

$\ln(x.y) = \ln x + \ln y$ $\ln(x/y) = \ln x - \ln y$

$\ln x^k = k.\ln x$ $\ln e^x = x$

</td>
</tr>
<tr>
<td>

$a^x = e^{x.\ln a}$

</td>
<td>

Iff $\ln g(x) = f(x)$ => $g(x) = e^{f(x)}$

</td>
</tr>
</table>

B. Properties of trigonometric functions

$$\cos^2\theta + \sin^2\theta = 1 \qquad \csc^2\theta = 1 + \cot^2\theta \qquad \sec^2\theta = 1 + \tan^2\theta$$

$$\sin(2\theta) = 2.\sin\theta.\cos\theta \quad \left| \quad \sin^2\theta = \frac{1 - \cos(2\theta)}{2} \right| \quad \cos^2\theta = \frac{1 + \cos(2\theta)}{2}$$

$$\cos(a+b) = \cos a.\cos b - \sin a.\sin b \qquad \cos(-b) = \cos b$$
$$\sin(a+b) = \sin a.\cos b + \cos a.\sin b \qquad \sin(-b) = -\sin b$$
$$\tan(a+b) = [\tan a + \tan b]/[1 - \tan a.\tan b] \qquad \tan(-b) = -\tan b$$

C. Properties of hyperbolic functions

$$\cosh^2 x - \sinh^2 x = 1 \qquad \text{csch}^2 x = \coth^2 x - 1 \qquad \text{sech}^2 t = 1 - \tanh^2 t$$

$$\sinh(2x) = 2.\sinh x.\cosh x \quad \left| \quad \sinh^2 x = \frac{\cosh(2x) - 1}{2} \right| \quad \cosh^2 x = \frac{\cosh(2x) + 1}{2}$$

$$\cosh(a+b) = \cosh a.\cosh b + \sinh a.\sinh b \qquad \cosh(-b) = \cosh b$$
$$\sinh(a+b) = \sinh a.\cosh b + \cosh a.\sinh b \qquad \sinh(-b) = -\sinh b$$
$$\tanh(a+b) = [\tanh a + \tanh b]/[1 + \tanh a.\tanh b] \qquad \tanh(-b) = -\tanh b$$

<table>
<tr>
<td>

$\cosh x = \dfrac{e^x + e^{-x}}{2}$

</td>
<td>

$\sinh x = \dfrac{e^x - e^{-x}}{2}$

</td>
<td>

$\cosh x + \sinh x = e^x$
$\cosh x - \sinh x = e^{-x}$

</td>
<td>

The even part of e^x is the $\cosh x$ and the odd part is the $\sinh x$

</td>
</tr>
</table>

Table of Standard Integrals

I. Rational powers of f(x)

$$\int f^p(x).df(x) = \frac{f^{p+1}(x)}{p+1} + C \quad p\#-1 \qquad \bigg| \qquad \int \frac{df(x)}{f(x)} = \int \frac{f'(x)}{f(x)}.dx = \ln|f(x)| + C \quad f(x)\#0$$

II. Exponential functions

$$\int e^{k.x}.dx = \frac{1}{k}.e^{k.x} + C \qquad \bigg| \qquad \int a^{k.x}.dx = \frac{1}{k.\ln a}.a^{k.x} + C \quad \begin{array}{l} a>0 \\ a\#1 \end{array}$$

III. Trigonometric functions

$\int \sin x.dx = -\cos x + C$	$\int \cos x.dx = \sin x + C$
$\int \tan x.dx = \ln|\sec x| + C$	$\int \cot x.dx = -\ln|\csc x| + C$
$\int \sec x.dx = \ln|\sec x+\tan x| + C$	$\int \csc x.dx = -\ln|\csc x+\cot x| + C$
$\int \sec^2 x.dx = \tan x + C \quad x\#(2k+1).\pi/2$	$\int \csc^2 x.dx = -\cot x + C \quad x\#k.\pi$

$$\int \csc u.\sec u.du = \ln|\tan u| + C$$

IV. Inverse trigonometric functions

$$\int \frac{dx}{\sqrt{1-x^2}} = \begin{array}{l} \arcsin x + C \\ -\arccos x + C \end{array} \quad |x|<1 \qquad \bigg| \qquad \int \frac{dx}{1+x^2} = \begin{array}{l} \arctan x + C \\ -\text{arccot} x + C \end{array}$$

V. Hyperbolic functions

$\int \sinh x.dx = \cosh x + C$	$\int \cosh x.dx = \sinh x + C$
$\int \tanh x.dx = -\ln|\text{sech} x| + C$	$\int \coth x.dx = -\ln|\text{csch} x| + C \quad x\#0$
$\int \text{sech}^2 x.dx = \tanh x + C$	$\int \text{csch}^2 x.dx = -\coth x + C \quad x\#0$
$\int \text{sech} x.dx = \arctan \sinh x + C$	$\int \text{csch} x.dx = -\ln|\text{csch} x+\coth x| + C$

VI. Inverse hyperbolic functions

$$\int \frac{dx}{\sqrt{x^2+1}} = \text{argsinh} x +C \quad\bigg|\quad \int \frac{dx}{\sqrt{x^2-1}} = \text{argcosh} x +C \quad\bigg|\quad \int \frac{dx}{1-x^2} = \begin{array}{ll} \text{argtanh} x +C & \text{argcoth} x +C \\ |x|<1 & |x|>1 \end{array}$$

Trigonometric reduction formulas

$u = \arcsin(x/a)$	$u = \text{arc csc}(x/a)$	$u = \arctan(x/a)$
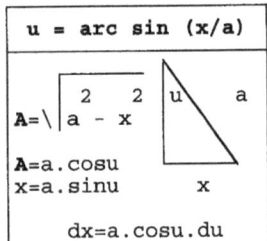	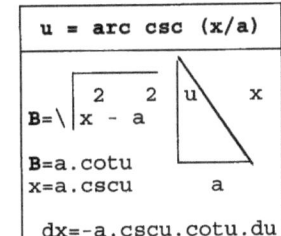	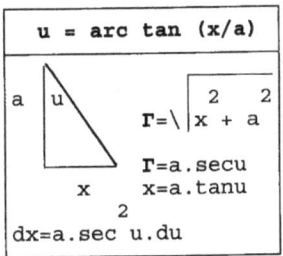
$A=\sqrt{a^2-x^2}$	$B=\sqrt{x^2-a^2}$	$\Gamma=\sqrt{x^2+a^2}$
$A=a.\cos u$ $x=a.\sin u$	$B=a.\cot u$ $x=a.\csc u$	$\Gamma=a.\sec u$ $x=a.\tan u$
$dx=a.\cos u.du$	$dx=-a.\csc u.\cot u.du$	$dx=a.\sec^2 u.du$

**For high powers of trigonometric integrals
the reduction formulas should be consulted**

1. $\displaystyle\int \sin^n\theta.d\theta = -\frac{\cos\theta.\sin^{n-1}\theta}{n} + \frac{n-1}{n}.\int \sin^{n-2}\theta.d\theta \qquad n>/2$

---

2. $\displaystyle\int \cos^n\theta.d\theta = \frac{\sin\theta.\cos^{n-1}\theta}{n} + \frac{n-1}{n}.\int \cos^{n-2}\theta.d\theta \qquad n>/2$

---

3. $\displaystyle\int \tan^n\theta.d\theta = \frac{\tan^{n-1}\theta}{n-1} - \int \tan^{n-2}\theta.d\theta \qquad n>/2$

---

4. $\displaystyle\int \cot^n\theta.d\theta = -\frac{\cot^{n-1}\theta}{n-1} - \int \cot^{n-2}\theta.d\theta \qquad n>/2$

---

5. $\displaystyle\int \sec^n\theta.d\theta = \frac{\tan\theta.\sec^{n-2}\theta}{n-1} + \frac{n-2}{n-1}.\int \sec^{n-2}\theta.d\theta \qquad n>/2$

---

6. $\displaystyle\int \csc^n\theta.d\theta = -\frac{\cot\theta.\csc^{n-2}\theta}{n-1} + \frac{n-2}{n-1}.\int \csc^{n-2}\theta.d\theta \qquad n>/2$

---

7. $\displaystyle\int \csc^n\theta.\sec\theta.d\theta = -\frac{\csc^{n-1}\theta}{n-1} + \int \csc^{n-2}\theta.\sec\theta.d\theta \qquad n>/2$

$$8. \int \sec^n\theta \cdot \csc\theta \cdot d\theta = \frac{\sec^{n-1}\theta}{n-1} + \int \sec^{n-2}\theta \cdot \csc\theta \cdot d\theta \qquad n > /2$$

$$9. \int \tan^n\theta \cdot \csc\theta \cdot d\theta = \frac{\csc\theta \cdot \tan^{n-1}\theta}{n-1} - \frac{n-2}{n-1} \cdot \int \tan^{n-2}\theta \cdot \csc\theta \cdot d\theta \qquad n > /2$$

$$10. \int \cot^n\theta \cdot \csc\theta \cdot d\theta = - \frac{\csc\theta \cdot \cot^{n-1}\theta}{n} - \frac{n-1}{n} \cdot \int \cot^{n-2}\theta \cdot \csc\theta \cdot d\theta \qquad n > /2$$

$$11. \int \cot^n\theta \cdot \sec\theta \cdot d\theta = - \frac{\sec\theta \cdot \cot^{n-1}\theta}{n-1} - \frac{n-2}{n-1} \cdot \int \cot^{n-2}\theta \cdot \sec\theta \cdot d\theta \qquad n > /2$$

$$12. \int \cos^n\theta \cdot \csc\theta \cdot d\theta = \frac{1}{n-1} \cdot \cos^{n-1}\theta + \int \cos^{n-2}\theta \cdot \csc\theta \cdot d\theta \qquad n > /2$$

$$13. \int \sin^n\theta \cdot \sec\theta \cdot d\theta = - \frac{1}{n-1} \cdot \sin^{n-1}\theta + \int \sin^{n-2}\theta \cdot \sec\theta \cdot d\theta \qquad n > /2$$

- Check those formulas by differentiation -

Taking into consideration that:

$\sin\theta$	$\cos\theta$
$\cos\theta$	$-\sin\theta$
$\tan\theta$	$\sec^2\theta$

$\csc\theta$	$-\csc\theta \cdot \cot\theta$
$\sec\theta$	$\sec\theta \cdot \tan\theta$
$\cot\theta$	$-\csc^2\theta$

$$\frac{d}{dx} f^p = p \cdot f^{p-1} \cdot f'$$

For example: The case 13. Just above:

$$\sin^n\theta \cdot \sec\theta = - \frac{n-1}{n-1} \cdot \sin^{n-2}\theta \cdot \cos\theta + \sin^{n-2}\theta \cdot \sec\theta$$

$$= \sin^{n-2}\theta \cdot \left[- \cos\theta + \sec\theta \right] = \sin^{n-2}\theta \cdot \left[1 - \cos^2\theta \right] \cdot \sec\theta$$

A list of integrals solved by trigonometric substitution

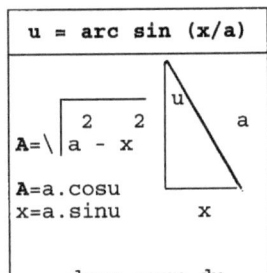

u = arc sin (x/a)

$A = \sqrt{a^2 - x^2}$

$A = a.\cos u$
$x = a.\sin u$

$dx = a.\cos u.du$

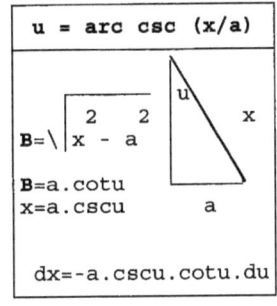

u = arc csc (x/a)

$B = \sqrt{x^2 - a^2}$

$B = a.\cot u$
$x = a.\csc u$

$dx = -a.\csc u.\cot u.du$

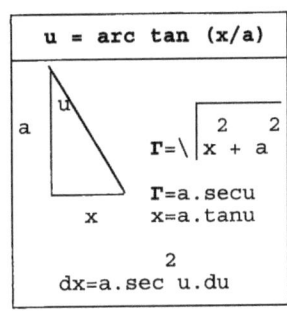

u = arc tan (x/a)

$\Gamma = \sqrt{x^2 + a^2}$

$\Gamma = a.\sec u$
$x = a.\tan u$

$dx = a.\sec^2 u.du$

$\displaystyle \int A^{-1}.dx = u \qquad + C$
$\qquad\qquad u = \text{arc sin } (x/a)$

$\displaystyle \int A^{-2}.dx = \frac{1}{2.a}.\ln\left|\frac{a+x}{a-x}\right| \quad + C$

$\displaystyle \int B^{-1}.dx = \ln|x+B| \qquad + C$

$\displaystyle \int B^{-2}.dx = \frac{1}{2.a}.\ln\left|\frac{x-a}{x+a}\right| \quad + C$

$\displaystyle \int \Gamma^{-1}.dx = \ln|x+\Gamma| \qquad + C$

$\displaystyle \int \Gamma^{-2}.dx = \frac{1}{a}.u \qquad + C$
$\qquad\qquad u = \text{arc tan } (x/a)$

$u = \text{arc sin } (x/a)$

$\displaystyle \int A.dx = \frac{1}{2}.\left[x.A + a^2.u\right] \qquad + C$

$\displaystyle \int \frac{A}{x}.dx = A - a.\ln\left|\frac{a+A}{x}\right| \qquad + C$

$\displaystyle \int B.dx = \frac{1}{2}.\left[x.B - a^2.\ln|x+B|\right] +C$

$\displaystyle \int \frac{B}{x}.dx = B + a.u \qquad + C$
$\qquad\qquad u = \text{arc csc } (x/a)$

$\displaystyle \int \Gamma.dx = \frac{1}{2}.\left[x.\Gamma + a^2.\ln|x+\Gamma|\right] +C$

$\displaystyle \int \frac{\Gamma}{x}.dx = \Gamma - a.\ln\left|\frac{a+\Gamma}{x}\right| \qquad + C$

$$\int A^{-1}.x^{-1}.dx = -a^{-1}.\ln\left|\frac{A+a}{x}\right| + C \quad\bigg|\quad \int A^{-2}.x^{-1}.dx = a^{-2}.\ln\left|\frac{x}{A}\right| + C$$

$$\int B^{-1}.x^{-1}.dx = -a^{-1}.[\text{arc csc }(x/a)] + C \quad\bigg|\quad \int B^{-2}.x^{-1}.dx = -a^{-2}.\ln\left|\frac{x}{B}\right| + C$$

$$\int \Gamma^{-1}.x^{-1}.dx = -a^{-1}.\ln\left|\frac{\Gamma+a}{x}\right| + C \quad\bigg|\quad \int \Gamma^{-2}.x^{-1}.dx = a^{-2}.\ln\left|\frac{x}{\Gamma}\right| + C$$

$$\int A^{-3}.dx = a^{-2}.\frac{x}{A} + C \quad\bigg|\quad \int A^{-4}.dx = \frac{1}{2}.a^{-3}.\left[\frac{x}{A}.\frac{a}{A} + \ln\left|\frac{a+x}{A}\right|\right] + C$$

$$\int B^{-3}.dx = -a^{-2}.\frac{x}{B} + C \quad\bigg|\quad \int B^{-4}.dx = -\frac{1}{2}.a^{-3}.\left[\frac{x}{B}.\frac{a}{B} - \ln\left|\frac{a+x}{B}\right|\right] + C$$

$$\int \Gamma^{-3}.dx = a^{-2}.\frac{x}{\Gamma} + C \quad\bigg|\quad \int \Gamma^{-4}.dx = \frac{1}{2}.a^{-3}.\left[\frac{x}{\Gamma}.\frac{a}{\Gamma} + \text{arc tan }(x/a)\right] + C$$

$$\int u.dx = x.u + A \qquad + C \quad\bigg|\quad \int v.dx = x.v - A \qquad + C$$

u=arc sin (x/a) | v=arc cos (x/a)

$$\int u.dx = x.u + a.\ln|x+B| \quad + C \quad\bigg|\quad \int v.dx = x.v - a.\ln|x+B| \quad + C$$

u=arc csc (x/a) | v=arc sec (x/a)

$$\int u.dx = x.u - a.\ln|\Gamma| \quad + C \quad\bigg|\quad \int v.dx = x.v + a.\ln|\Gamma| \quad + C$$

u=arc tan (x/a) | v=arc cot (x/a)

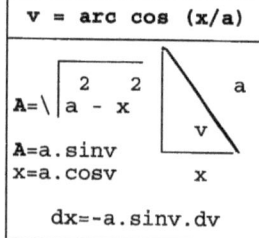

v = arc cos (x/a)

$A=\sqrt{a^2 - x^2}$

A=a.sinv
x=a.cosv

dx=-a.sinv.dv

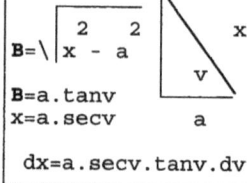

v = arc sec (x/a)

$B=\sqrt{x^2 - a^2}$

B=a.tanv
x=a.secv

dx=a.secv.tanv.dv

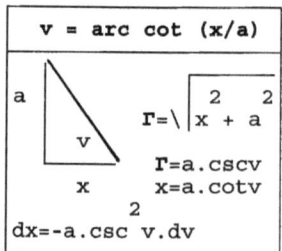

v = arc cot (x/a)

$\Gamma=\sqrt{x^2 + a^2}$

Γ=a.cscv
x=a.cotv

dx=-a.csc²v.dv

Hyperbolic integrals solved by trigonometric substitution

$\int \sinh x \, dx = \cosh x + C$	$\int \cosh x \, dx = \sinh x + C$
$\int \tanh x \, dx = -\ln\lvert \operatorname{sech} x \rvert + C$	$\int \coth x \, dx = -\ln\lvert \operatorname{csch} x \rvert + C \quad x \neq 0$
$\int \operatorname{sech}^2 x \, dx = \tanh x + C$	$\int \operatorname{csch}^2 x \, dx = -\coth x + C \qquad x \neq 0$
$\int \operatorname{sech} x \, dx = \arctan \sinh x + C$	$\int \operatorname{csch} x \, dx = -\ln\lvert \operatorname{csch} x + \coth x \rvert + C$

$$\tan u = \sinh x \qquad \frac{d}{du}\tan u = \sec^2 u \qquad \frac{d}{dx}\sinh x = \cosh x$$

$$\sec^2 u \, . du = \cosh x \, . dx = \sec u \, . dx \quad \Rightarrow \quad \boxed{\sec u \, . du = dx}$$

Category I.
Not exclusively solved by trigonometric substitution

$$\int \sinh x \, dx = \int \tan u \, . [\sec u] \, . du = \int d\sec u = \cosh x + C$$

$$\int \cosh x \, dx = \int \sec u \, . [\sec u] \, . du = \int d\tan u = \sinh x + C$$

$$\int \tanh x \, dx = \int \tan u \, . \cos u \, . [\sec u] \, . du = \int d\ln\sec u = \ln\lvert \cosh x \rvert + C$$
$$= -\ln\lvert \operatorname{sech} x \rvert + C$$

$$\int \coth x \, dx = \int \sec u \, . \cot u \, . [\sec u] \, . du = \int d\ln\tan u = \ln\lvert \sinh x \rvert + C$$
$$= \int \csc u \, . [\sec u] \, . du \qquad\qquad = -\ln\lvert \operatorname{csch} x \rvert + C$$

Category II.
Exclusively solved by trigonometric substitution

$$\int \operatorname{sech} x \, dx = \int \frac{1}{\cosh x} \, . [\sec u] \, . du = \int \frac{1}{\sec u} \, . [\sec u] \, . du = \int du$$

$$u = \operatorname{arcsec} \cosh x = \arctan \sinh x$$

$r = (a + i.b)$

The integral $\quad In = \int e^{r.x} . x^n . dx \quad$ **solved** $\quad$ **by trigonometric substitution**

n = any non negative integer n $\quad \square \quad e^{r.x} = e^{t}$

$$\int e^{r.x} . x^n . dx = r^{-(n+1)} . \int e^{r.x} . (r.x)^n . d(r.x) = r^{-(n+1)} . \int e^{t} . t^n . dt$$

$$e^{t} = \cosh t + \sinh t = \sec u + \tan u$$

$$t = \ln[\sec u + \tan u]$$

$1 \quad$ (triangle with u, hypotenuse, $c = \cosh t = \sec u$)

$s = \sinh t = \tan u$

$$dt = \frac{\sec u . \tan u + \sec^2 u}{\sec u + \tan u} . du = \sec u . du$$

$$\int e^{t} . t^n . dt = \int [\sec u + \tan u] . \ln^n [\sec u + \tan u] . \sec u . du \quad = e^{t}$$

$$[\sec u + \tan u] = \xi$$

$$= \int \ln^n [\sec u + \tan u] . d[\tan u + \sec u]$$

- -

$$= \int \ln^n \xi . d\xi = \xi . \left[\ln^n \xi - n.\ln^{n-1} \xi + n.(n-1).\ln^{n-2} \xi \right.$$
$$- n.(n-1).(n-2).\ln^{n-3} \xi$$
$$+ n.(n-1).(n-2).(n-3).\ln^{n-4} \xi$$
$$- n.(n-1).(n-2).(n-3).(n-4).\ln^{n-5} \xi$$
$$+ \ldots \ldots \ldots \ldots \ldots \ldots \ldots$$
$$\left. + (-1)^n . n! \right]$$

Something | **like** -

succesive derivatives with alternating sign

- -

$$= e^{r.x} . \left[(r.x)^n - n.(r.x)^{n-1} + \ldots + (-1)^n . n! \right] + C$$

And: $In = \int e^{r.x} . x^n . dx =$

$$= e^{r.x} . \left[x^n . r^{-1} - n.x^{n-1} . r^{-2} + n.(n-1).x^{n-2} . r^{-3} - \ldots + (-1)^n . n! . r^{-(n+1)} \right] + C$$

Algebraic trasformations applied to quadratic trinomial suitable for integration

$$p2(x) = a.x^2 + b.x + c = a.\left[\underbrace{x + \frac{b}{2.a}}_{=t}\right]^2 + \frac{4.a.c - b^2}{4.a^2} \qquad \text{Taylor's series expansion}$$

$$= a.\left[t^2 + k^2\right] \quad k = \frac{\sqrt{-\Delta}}{2.a} \quad \begin{matrix}\Delta<0\\a>0\end{matrix} \quad \boxed{\text{Or}} = a.\left[t^2 - \lambda^2\right] \quad \lambda = \frac{\sqrt{\Delta}}{2.a} \quad \begin{matrix}\Delta>0\\a>0\end{matrix} \quad \boxed{\Delta = b^2 - 4ac}$$

$$\boxed{a>0} \quad \int \frac{dx}{p2(x)} = \frac{1}{a}.\frac{1}{k}.\arctan\left(\frac{t}{k}\right) + C \quad \boxed{\text{Or}} = \frac{1}{a}.\frac{1}{2.\lambda}.\ln\left|\frac{t-\lambda}{t+\lambda}\right| + C$$
$$\qquad\qquad\qquad\qquad\qquad\qquad \Delta<0 \qquad\qquad\qquad\qquad\qquad\qquad\qquad \Delta>0$$

$$a>0 \quad \frac{1}{x.p2(x)} = \frac{1}{c}.\left[\frac{1}{x} - \frac{a.x+b}{p2(x)}\right] = \frac{1}{c}.\left[\frac{1}{x} - \frac{1}{2}.\frac{p2'(x)}{p2(x)} - \frac{b}{2}.\frac{1}{p2(x)}\right]$$

1st and 2nd term evaluated by the golden rule for integrals:

$$a>0 \quad \frac{1}{x.\left[p2(x)\right]^2} = \frac{1}{c}.\left[\frac{1}{x.p2(x)} - \frac{1}{2}.\frac{p2'(x)}{\left[p2(x)\right]^2} - \frac{b}{2}.\frac{1}{\left[p2(x)\right]^2}\right]_*$$

* See trigonometric reduction formulas:

$$a>0 \quad \frac{1}{x.\left[p2(x)\right]^3} = \frac{1}{c}.\left[\frac{1}{x.\left[p2(x)\right]^2} - \frac{1}{2}.\frac{p2'(x)}{\left[p2(x)\right]^3} - \frac{b}{2}.\frac{1}{\left[p2(x)\right]^3}\right]_* \text{etc}$$

$$a>0 \quad \frac{1}{x^2.p2(x)} = \frac{1}{c}.\left[\frac{1}{x^2} - \frac{b}{x^1.p2(x)} - \frac{a}{x^0.p2(x)}\right] \qquad x^0 = 1$$

$$a>0 \quad \frac{1}{x^3.p2(x)} = \frac{1}{c}.\left[\frac{1}{x^3} - \frac{b}{x^2.p2(x)} - \frac{a}{x^1.p2(x)}\right] \qquad \text{etc}$$

- A golden rule for integrals -
often not recognized

Look always at the integrand function to see whether it contains (or can be made to contain) **its 1st derivative**

- Iff it contains this derivative the solution is instant -

$$\int f^p(x).f'(x).dx = \frac{f^{p+1}(x)}{p+1} + C \quad \bigg| \quad \int f^p(x).f'(x).dx = \ln|f(x)| + C$$

p#-1 (left)   p=-1 (right)

Valid also for p=0

Examples:

1.
$$\int x^k.dx = \frac{x^{k+1}}{k+1} + C \qquad k\#-1$$

2.
$$\int \frac{dx}{ax+b} = \frac{1}{a}.\ln(ax+b) + C$$

3.
$$\int \frac{dx}{1-x} = -\ln[1-x] + C$$

4.
$$\int a^{k.x}.dx = \frac{1}{r.k}.\int r.k.a^{k.x}.dx = \frac{1}{r.k}.\int da = \frac{1}{r.k}.a^{k.x} + C \quad \bigg| \quad a^{k.x} = e^{r.k.x}$$

a>0   a#1   r=lna

5.
$$\int \frac{dx}{\sqrt{ax+b}} = \frac{2}{a}.\sqrt{ax+b} + C$$

f'(x)=a

6.
$$\int \sqrt{ax+b}.dx = \frac{2}{3.a}.\sqrt{(ax+b)^3} + C$$

7.
$$\int \sqrt[n]{ax+b}.dx = \frac{2.(ax+b)}{(n+2).a}.\sqrt[n]{ax+b} + C$$

8.
$$\int x.\sqrt{x^2 + a^2}.dx = \frac{1}{3}.(x^2 + a^2)^{3/2} + C$$

9.
$$\int \frac{x}{x^2 + a^2}.dx = \ln\left[\sqrt{x^2 + a^2}\right] + C$$

$$\frac{1}{2}.\int f^{1/2}.f'.dx = \frac{1}{2}.\frac{f^{1/2+1}}{1/2+1} + C$$

$$\frac{1}{2}.\int f^{-1}.f'.dx = \frac{1}{2}.\ln|f| + C$$

Trigonometric transformations suitable for integration

Multiplication of nominator and denominator by $\sec^2\left(\dfrac{a.x}{2}\right)$

$$\int \frac{dx}{b + c.\cos(a.x)} = \frac{2}{a.(b-c)} \cdot \int \frac{dt}{t^2 + \dfrac{b+c}{b-c}} \qquad t=\tan\left(\frac{a.x}{2}\right)$$

$$\int \frac{dx}{b + c.\sin(a.x)} = \frac{2}{a.b} \cdot \int \frac{d\xi}{\xi^2 + (1-\mu^2)} \qquad \boxed{\xi=(t+\mu)} \quad t=\tan\left(\frac{a.x}{2}\right) \quad \mu=\left(\frac{c}{b}\right)$$

$$\textbf{(22.a) \& (22.b)} \quad \int_{O}^{\pi/2} \frac{dx}{1+c.\cos x} = \int_{O}^{\pi/2} \frac{dx}{1+c.\sin x} = \frac{\text{arc cos c}}{\sqrt{1-c^2}}$$

$$|c|<1$$

The numbers in front of the integrals are the corresponding
numbers of the table of 58 definite integrals

$$\textbf{(23.)} \int_{O}^{\pi} \frac{dx}{1+c.\sin x} = \frac{2.(\text{arc cos c})}{\sqrt{1-c^2}} \qquad \textbf{(24.)} \int_{O}^{\pi} \frac{dx}{1+c.\cos x} = \frac{\pi}{\sqrt{1-c^2}}$$

$$(-1<c<1) \qquad\qquad (-1<c<1)$$

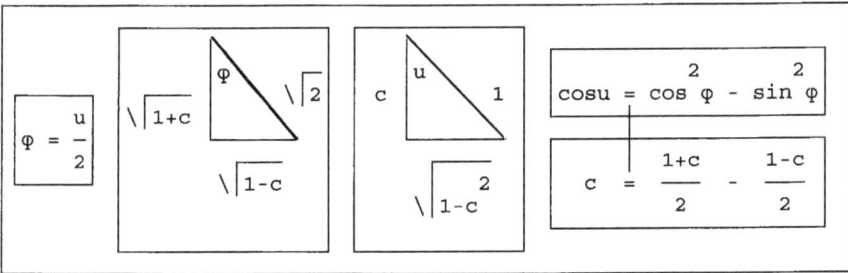

Trigonometric integrals solved by trigonometric substitution

$$\sin(m.x).\sin(n.x) = (1/2).\{ \cos[(m-n).x] - \cos[(m+n).x] \}$$

$$\cos(m.x).\cos(n.x) = (1/2).\{ \cos[(m-n).x] + \cos[(m+n).x] \}$$

$$\sin(m.x).\cos(n.x) = (1/2).\{ \sin[(m-n).x] + \sin[(m+n).x] \}$$

$$\int^{m\neq n} \sin(m.x).\sin(n.x).dx = \frac{\sin[(m-n).x]}{2.(m-n)} - \frac{\sin[(m+n).x]}{2.(m+n)} + C$$

- -

$$\int^{m\neq n} \cos(m.x).\cos(n.x).dx = \frac{\sin[(m-n).x]}{2.(m-n)} + \frac{\sin[(m+n).x]}{2.(m+n)} + C$$

- -

$$\int^{m\neq n} \sin(m.x).\cos(n.x).dx = -\frac{\cos[(m-n).x]}{2.(m-n)} - \frac{\cos[(m+n).x]}{2.(m+n)} + C$$

---

29. $\displaystyle\int_0^\pi \sin(m.x).\sin(n.x).dx$ $\Big|$ $= 0$ (m≠n integers)

$= \pi/2$ (m=n integers)

---

30. $\displaystyle\int_0^\pi \cos(m.x).\cos(n.x).dx$ $\Big|$ $= 0$ (m≠n integers)

$= \pi/2$ (m=n≠0 integers) $\boxed{=\pi}$ (m=n=0)

---

31. $\displaystyle\int_0^\pi{}^{m\neq n} \sin(m.x).\cos(n.x).dx$ $\Big|$ $= 0$ (m,n integers , m+n **even**)

$= \dfrac{2.m}{m^2 - n^2}$ (m,n integers , m+n **odd**)

$$\int_0^\lambda \sin(um.x).\sin(un.x).dx \;\Big|\; = 0 \quad (m\neq n \text{ integers})$$

$= \lambda/2$ (m=n integers)

$$\int_0^\lambda \cos(um.x).\cos(un.x).dx \;\Big|\; = 0 \quad (m\neq n \text{ integers})$$

$= \lambda/2$ (m=n≠0 integers) $\boxed{=\lambda}$ (m=n=0)

$$\int_0^\lambda{}^{m\neq n} \sin(um.x).\cos(un.x).dx \;\Big|\; = 0 \quad (m,n \text{ integers}, m+n \text{ even})$$

$um=m.\dfrac{\pi}{\lambda} \qquad un=n.\dfrac{\pi}{\lambda}$

$= \dfrac{\lambda}{\pi}.\dfrac{2.m}{m^2 - n^2}$ (m,n integers , m+n **odd**)

That's why I am fond of Trigonometry

The ancient Greeks said: Αεί ο Θεός ο Μέγας γεωμετρεί
3 , 1   4   1   5   9

Estimation of the number π by Archimedes

The Great God constructing this world continuously
applied the rules of geometry

I would add the rules of geometry and trigonometry
which converts geometry into algebra

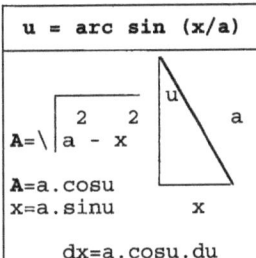

$u = \text{arc sin } (x/a)$

$A = \sqrt{a^2 - x^2}$

$A = a.\cos u$
$x = a.\sin u$

$dx = a.\cos u.du$

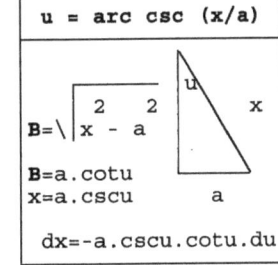

$u = \text{arc csc } (x/a)$

$B = \sqrt{x^2 - a^2}$

$B = a.\cot u$
$x = a.\csc u$

$dx = -a.\csc u.\cot u.du$

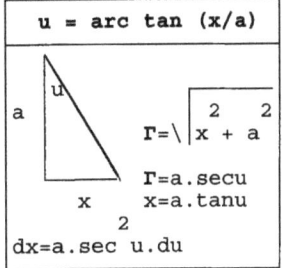

$u = \text{arc tan } (x/a)$

$\Gamma = \sqrt{x^2 + a^2}$

$\Gamma = a.\sec u$
$x = a.\tan u$

$dx = a.\sec^2 u.du$

Another proof that:

$$(13.) \quad \int_0^{+\infty} \frac{dx}{\left[a^2 + x^2\right]^n} = \frac{1}{a^{2n-1}} \cdot \left[\frac{(2n)!}{(2^n.n!)^2} \cdot \frac{\pi}{2}\right] \cdot \frac{2n}{2n-1} \qquad \begin{array}{l} a>0 \\ n=2,3,.. \end{array}$$

$$(13.) = \int_0^{\pi/2} (a^{-2}.\cos^2 u)^n .a.\sec^2 u.du = a^{-2n+1} \cdot \int_0^{\pi/2} \cos^{2n-2} u.du$$

$$= \frac{1}{2} \cdot \frac{1}{a^{2n-1}} \cdot \frac{\Gamma(n-1/2).\Gamma(1/2)}{\Gamma(n)} = \frac{1}{2} \cdot \frac{1}{a^{2n-1}} \cdot \frac{\Gamma(n+1/2).\Gamma(1/2)}{\Gamma(n+1)} \cdot \frac{2n}{2n-1}$$

The numbers in front of the integrals are the corresponding
numbers of the table of 58 definite integrals

$$B(x,y)$$

$$B(x,y) = \frac{\Gamma(x).\Gamma(y)}{\Gamma(x+y)} \qquad \boxed{x,y > 0}$$

$$= \int_0^1 (1-t)^{y-1}.t^{x-1}.dt = \int_0^{+\infty} \frac{u^{x-1}}{(1+u)^{x+y}}.du = 2.\int_0^{\pi/2} \sin^{2x-1}\theta.\cos^{2y-1}\theta.d\theta$$

Another way of prooving the equivalent forms of B(x,y)

Another form of confirming
what the ancient Greeks said:

$$\int_0^{+\infty} \frac{\xi^{x-1}}{(1+\xi)^{x+y}}.d\xi = 2.\int_0^{+\infty} \frac{(t^2)^{x-1}}{(1+t^2)^{x+y}}.t.dt \qquad 1 \underset{t=\tan u}{\overset{u}{\diagdown}} \boxed{\sqrt{1+t^2}=\sec u}$$

$$= 2.\int_0^{\pi/2} \sin^{2x-2}u.\sec^{2x-2}u.\cos^{2x+2y}u.\underbrace{\sin u.\sec u}_{=t}.\underbrace{\sec^2 u.du}_{=dt}$$

$$= 2.\int_0^{\pi/2} \sin^{2x-1}u.\cos^{2y-1}u.du = \boxed{I}$$

$$\int_0^1 (1-\xi)^{y-1}.\xi^{x-1}.d\xi = 2.\int_0^1 (1-t^2)^{y-1}.(t^2)^{x-1}.t.dt$$

$$= 2.\int_0^1 (\cos^2 u)^{y-1}.(\sin^2 u)^{x-1}.\sin u.d\sin u$$

$$\boxed{I} = 2.\int_0^{\pi/2} (\cos^2 u)^{y-1}.(\sin^2 u)^{x-1}.\sin u.\cos u.du$$

The Wallis product formula:

As $n \to +\infty$ $(n+\frac{1}{2}) \cdot \Gamma^2(n+\frac{1}{2}) = \Gamma^2(n+1)$ $\Rightarrow$ $\dfrac{\pi}{2} = \lim\limits_{n \to +\infty} \dfrac{[2^n \cdot n!]^4}{[(2n)!]^2 \cdot (2n+1)}$

Corollary: $\sqrt{\pi} = \Gamma(1/2) = \lim\limits_{n \to +\infty} \dfrac{(2^n \cdot n!)^2}{(2n)!} \cdot \dfrac{1}{\sqrt{n}}$

Wallis derived the formulas of K,L,M by the following trigonometric reduction formula

$$\int \sin^n \theta \cdot d\theta = -\frac{\cos\theta \cdot \sin^{n-1}\theta}{n} + \frac{n-1}{n} \cdot \int \sin^{n-2}\theta \cdot d\theta \qquad n > /2$$

In the interval of integration of K,L,M
- Only the even powers of $\sin\theta$ carry the π -

$\displaystyle\int_0^{\pi/2} \sin^{2n-1}\theta \cdot d\theta = \frac{1}{2} \cdot \frac{\Gamma(n) \cdot \Gamma(1/2)}{\Gamma(n+1/2)} = K$ | $\Gamma(n) = \dfrac{n!}{n}$ | $\Gamma(n+\frac{1}{2}) = \dfrac{1}{2^{2n}} \cdot \dfrac{(2n)!}{2^n \cdot n!} \cdot \sqrt{\pi}$

$\displaystyle\int_0^{\pi/2} \sin^{2n}\theta \cdot d\theta = \frac{1}{2} \cdot \frac{\Gamma(n+1/2) \cdot \Gamma(1/2)}{\Gamma(n+1)} = L$ | $\Gamma(n+1) = n!$ | $\Gamma(\frac{1}{2}) = \sqrt{\pi}$

$\displaystyle\int_0^{\pi/2} \sin^{2n+1}\theta \cdot d\theta = \frac{1}{2} \cdot \frac{\Gamma(n+1) \cdot \Gamma(1/2)}{\Gamma(n+3/2)} = M$ | $\Gamma(n+3/2) = (n+\frac{1}{2}) \cdot \Gamma(n+\frac{1}{2})$

As $n \to +\infty$ and to the limit: $\dfrac{M}{K} = \dfrac{2n}{2n+1} = 1$

Since $0 < \sin\theta < 1$ for $0 < \theta < \pi/2$

$\sin^{2n-1}\theta > \sin^{2n}\theta > \sin^{2n+1}\theta$ ie: $\boxed{0 < M < L < K}$

Inequalities of integrand functions implies inequalities of integrals over the same interval of integration (but not vice versa)

As $n \to +\infty$ and to the limit: $L = M \Rightarrow (n+\frac{1}{2}) \cdot \Gamma^2(n+\frac{1}{2}) = \Gamma^2(n+1)$

A unique approach to Fourier's series and integral

t is the variable , while x is fixed $(-\pi < x < \pi)$

$$\cos[k.(t-x)] = \cos(k.x).\cos(k.t) + \sin(k.x).\sin(k.t)$$

$$Sf(x) = \frac{1}{\pi} . \int_{-\pi}^{\pi} f(t) . \left[\frac{1}{2} + \sum_{k=1}^{+\infty} \cos[k.(t-x)] \right].dt = \frac{f(x-)+f(x+)}{2} \;^* = f(x)$$

* Iff f(x) is continuous at selected fixed x

$$Sf(x) = \frac{1}{\pi} . \int_{-\pi}^{\pi} f(t) . \left[\int_{0}^{+\infty} \cos[k.(t-x)].dk \right].dt$$

Evaluated by the average value over the unit interval
Discrete variable k converted into a continuous one

$$Sf(x) = \frac{1}{\pi} . \int_{-\pi}^{\pi} f(t) . \left[\frac{\sin[k.(t-x)]}{(t-x)} \Big|_{k=0}^{k+\infty} \right].dt \quad \boxed{\begin{array}{l}\text{By evaluation of}\\\text{the integral of dk}\end{array}}$$

$$= \frac{1}{\pi} . \int_{-\pi}^{\pi} f(t) . \frac{\sin[\omega.(t-x)]}{(t-x)}^{\omega->+\infty}.dt = \int_{-\pi}^{\pi} f(t).R(t-x).dt = \frac{f(x-)+f(x+)}{2}$$

$$Sf(x) = \frac{2}{\pi} . \int_{0}^{\pi} f(t) . \left[\int_{0}^{+\infty} \begin{array}{l} \cos(k.x).\cos(k.t) \\ \text{or} \\ \sin(k.x).\sin(k.t) \end{array} .dk \right].dt \quad \begin{array}{l} f(x) \textbf{ even} \\ \text{--} \\ f(x) \textbf{ odd} \end{array}$$

$$= \frac{2}{\pi} . \int_{0}^{+\infty} . \int_{0}^{+\infty} f(t) . \begin{array}{l} \cos(k.x).\cos(k.t) \\ \text{or} \\ \sin(k.x).\sin(k.t) \end{array} .dk.dt \quad \begin{array}{l} f(x) \textbf{ even} \\ \text{--.--} \\ f(x) \textbf{ odd} \end{array}$$

$$\text{or dt.dk}$$

Provided that: $\int_{-\infty}^{+\infty} |f(t)|.dt$ converges $\quad$ The same interval of integration

The series of delta functions P(t-x)

$$P(t-x) = \frac{1}{\pi}.\left[\frac{1}{2} + \sum_{k=1}^{+\infty} \cos[k.(t-x)] \right] = \frac{1}{\pi}.\left[\frac{\sin[\omega.(t-x)]}{2.\sin[(t-x)/2]} \right]^{\omega=->+\infty}$$

The delta function $R(t-x) = \frac{1}{\pi}.\left[\frac{\sin[\omega.(t-x)]}{t-x} \right]^{\omega->+\infty}$
A continuous function

The graph of P(x) for n=10 and the graph of the function R(x)

As n->+oo the peaks tend to +oo and the small vibrations become convergent to zero infinitesimals

$$P(x) = -\frac{1}{\pi}\left[\frac{1}{2} + \sum_{k=1}^{+oo} \cos(k.x)\right] = -\frac{1}{\pi}.\frac{\sin(\omega.x)}{2.\sin(x/2)} \quad \xrightarrow{\omega->+oo}$$

$$R(x) = -\frac{1}{\pi}.\frac{\sin(\omega.x)}{x} \quad \xrightarrow{\omega->+oo}$$

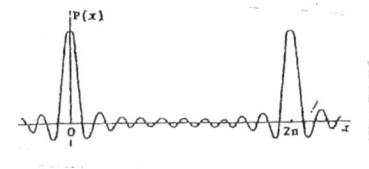

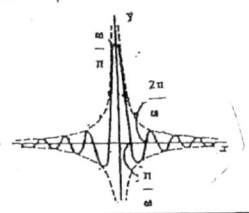

By translation of the axis we get the P(t-x) and R(t-x)

The heretic Dirac's isosceles triangle

. Extreme height -> +oo
. Very narrow base -> O
. At the middle of this base is the fixed x
. The total aerea of this triangle is the unity $1 = \frac{1}{2} + \frac{1}{2}$

.Outside of this very narrow base ie for t#x the Dirac's delta function , which is a continuous function , is an infinitesimal so that the product of this delta function by the function f(t) t#x is a **strong** infinitesimal , not contributing to the final outcome of the integral.

For any k#O

$$\int_{-\pi}^{\pi} \cos(k.t).dt = \frac{\sin(k.t)}{k}\bigg|_{k.t=-k.\pi}^{k.t=k.\pi} = 0 \qquad \int_{-\pi}^{\pi} \sin(k.t).dt = \frac{-\cos(k.t)}{k}\bigg|_{-k.\pi}^{k.\pi} = 0$$

Orthogonality properties in the given domain

$$\int_{-\pi}^{\pi} \cos^2(k.t).dt = \frac{1}{k}.\left[\frac{1}{2}.[k.t + \sin(k.t).\cos(k.t)]\right]\bigg|_{k.t=-k.\pi}^{k.t=k.\pi} = \pi$$

$$\int_{-\pi}^{\pi} \sin^2(k.t).dt = \frac{1}{k}.\left[\frac{1}{2}.[k.t - \sin(k.t).\cos(k.t)]\right]\bigg|_{k.t=-k.\pi}^{k.t=k.\pi} = \pi$$

Expansion in orthogonal functions
Complete set of
orthogonal functions
and the role of Bessel's equation

The $J_n(bnj.x)$ is like a sine with infinite solutions

ie: $J_n(bnj.x) = 0$

Replacing x by A (boundary condition) the $bnj^{j=1,2,3,..}$ are found

**The bni=yi and bnj=yj are determined from
the roots of the boundary condition**
- Frequently in physical problems K or L are zero -

$$K.J_n(bni.A) + L.\frac{d}{dx} J_n(bni.A) = 0 \qquad \Big| \qquad K.J_n(bnj.A) + L.\frac{d}{dx} J_n(bnj.A) = 0$$

$$\left[f(x) - \sum_{j=1}^{+\infty} A_j.J_n(bnj.x) \right]$$

----=Sf(x)-----

As small as possible
Complete set of orthogonal functions
t is the variable , while x is fixed
$t_1 < x < t_2$

$$Sf(x) = \int_{t_1=0}^{t_2=A} f(t). \left[\sum_{j=1}^{+\infty} \frac{J_n(bnj.x).J_n(bnj.t)}{N_j} \right].t.dt = \frac{f(x-)+f(x+)}{2}$$

By evaluation of the integral

Behaves like a $\delta\omega(t-x)$ $\omega(t)=t$

The heretic Dirac's isosceles triangle
Reference 1. (Appendix)
* Surprize

. Extreme height -> $+\infty$
. Very narrow base -> O
. At the middle of this base is the fixed x
. The total aerea of this triangle is the unity $1 = \frac{1}{2} + \frac{1}{2}$

.Outside of this very narrow base ie for t#x the Dirac's delta
function , which is a continuous function , is an infinitesimal
so that the product of this delta function by the function f(t)
t#x is a **strong** infinitesimal , not contributing to the final
outcome of the integral.

* Most of the classical textbooks avoid to mention the delta
function,since it is considered as a heresy of mathematics

Suppose that the function f(x) has a finite number
of discontinuities in the domain: $O=x1< x < x2=A$

$$\int_0^A Jn(bnj.x).Jn(bni.x).x.dx = x. \left. \frac{\begin{vmatrix} Jn(bnj.x) & [d/dx \ Jn(bnj.x)] \\ Jn(bni.x) & [d/dx \ Jn(bni.x)] \end{vmatrix}}{(bnj^2 - bni^2)} \right|_{x=0}^{x=A} = O$$

bnj#bni

-For the roots of either the function or of its 1st derivative.-

But not for exactly the same roots since bnj#bni

$$\int_0^A Jn^2(bnj.x).x.dx = Nj$$

$$\int_0^A Jn^2(bnj.x).x.dx \underset{bnj=r}{=} \frac{1}{2.r^2} . \left. \left[x^2 . \left[\frac{d}{dx}Jn(r.x)\right]^2 + \left[r^2.x^2 -n^2\right].Jn^2(r.x) \right] \right|_{x=0}^{x=A}$$

$$= Nj \qquad \text{- For exactly the same roots -}$$

Orthogonality: $\displaystyle \int_0^A Jn(bnj.x).Jn(bni.x).x.dx = \begin{matrix} O & i\#j \\ Nj & i=j \end{matrix}$

And:

$$Sf(x) = \int_{t1=0}^{t2=A} f(t). \left[\sum_{j=1}^{+\infty} \frac{Jn(bnj.x).Jn(bnj.t)}{Nj} \right].t.dt = \frac{f(x-)+f(x+)}{2}$$

By evaluation
of the integral   Behaves like a $\delta\omega(t-x)$ $\omega(t)=t$

Reference for the Dirac's heretic isosceles triangle

Higher Mathematics for beginners:
by Ya.B.Zeldovich
(Mir Publishers Moscow 1973)

Elliptic integrals

The solution of the **elliptic integrals** requires binomial expansion of the integrand function and the application of the beta function for the even powers of sinθ.

$$x = e^2 . \sin^2 \theta \qquad e<1$$

$$Ia = \int_0^{\pi/2} \sqrt{1-e^2 . \sin^2 \theta} \; .d\theta$$

$$e<1$$

$$Ia = \frac{\pi}{2} . \left[1 - \sum_{n=1}^{+\infty} \frac{[(2n)!]^2}{(2^n . n!)^4 . (2n-1)} . e^{2n} \right]$$

$$x = k^2 . \sin^2 \theta \qquad k<1$$

$$Ib = \int_0^{\pi/2} \frac{d\theta}{\sqrt{1-k^2 . \sin^2 \theta}}$$

$$k<1$$

$$Ib = \frac{\pi}{2} . \left[1 + \sum_{n=1}^{+\infty} \frac{[(2n)!]^2}{(2^n . n!)^4} . k^{2n} \right]$$

$$\sqrt{1-x} = 1 - \frac{1}{2} . [x/1] - \frac{1.3}{2.4} . [x^2 /3] - \frac{1.3.5}{2.4.6} . [x^3 /5] - \frac{1.3.5.7}{2.4.6.8} . [x^4 /7]$$

$$= 1 - \sum_{n=1}^{+\infty} \frac{(2n)!}{(2^n . n!)^2 . (2n-1)} . x^n \qquad - \frac{1.3.5.7.9}{2.4.6.8.10} . [x^5 /9] - ..$$

$$|x|<1$$

Ratio of odd factorials **up to 2n-3** by even factorials up to 2n

---

$$\frac{1}{\sqrt{1-x}} = 1 + \frac{1}{2} . x + \frac{1.3}{2.4} . x^2 + \frac{1.3.5}{2.4.6} . x^3 + \frac{1.3.5.7}{2.4.6.8} . x^4$$

$$= 1 + \sum_{n=1}^{+\infty} \frac{(2n)!}{(2^n . n!)^2} . x^n \qquad + \frac{1.3.5.7.9}{2.4.6.8.10} . x^5 + ..$$

$$|x|<1$$

Ratio of odd factorials up to 2n-1 by even factorials up to 2n

---

$$\int_0^{\pi/2} \sin^{2n} \theta .d\theta = \frac{1}{2} . \frac{\Gamma[n+(1/2)] . \Gamma(1/2)}{\Gamma(n+1)} = \frac{(2n)!}{(2^n . n!)^2} . \frac{\pi}{2}$$

Two cases of elliptic integrals

1. Element of arc length of vertical ellipse:

$$(x/a)^2 + (y/b)^2 = 1 \quad (b>a)$$

$$dl = \sqrt{(dx)^2 + (dy)^2} \overset{b>a}{=} b.\sqrt{1-e^2.\sin^2 t}\,.dt \qquad \boxed{e=(c/b)<1} \qquad x=a.\sin t \quad y=b.\sin t \qquad c=\sqrt{b^2-a^2}$$

e = eccentricity of the vertical ellipse

$$Ia = \int_0^{\pi/2}\sqrt{1-e^2.\sin^2\theta}\,.d\theta = \frac{\pi}{2}.\left[1 - \sum_{n=1}^{+\infty}\frac{[(2n)!]^2}{(2^n.n!)^4.(2n-1)}.e^{2n}\right] \qquad e^2<1$$

Total arclength L of the vertical ellipse

$$L = 2\pi.b.\left[1 - \left[\frac{1}{2}\right]^2.e^2 - \left[\frac{1.3}{2.4}\right]^2.\frac{e^4}{3} - \left[\frac{1.3.5}{2.4.6}\right]^2.\frac{e^6}{5} - ...\right]$$

----

2. The period T of the mathematical pendulum

$$T = 4.\sqrt{\frac{L}{2.g}}.\int_0^{a=\varphi_0}\frac{d\varphi}{\sqrt{\cos\varphi-\cos a}} = 4.\sqrt{\frac{L}{g}}.\int_0^{\pi/2}\frac{d\theta}{\sqrt{1-k^2.\sin^2\theta}} \qquad \text{An elliptic integral}$$

$$0<\varphi<\varphi_0=a \qquad\qquad\qquad k=\sin(a/2) \quad \varphi_0=a$$

$$Ib = \int_0^{\pi/2}\frac{d\theta}{\sqrt{1-k^2.\sin^2\theta}} = \frac{\pi}{2}.\left[1 + \sum_{n=1}^{+\infty}\frac{[(2n)!]^2}{(2^n.n!)^4}.k^{2n}\right] \qquad k^2<1$$

-See next page-

$$T = 2\pi.\sqrt{L/g}.\left[1 + \left[\frac{1}{2}\right]^2.k^2 + \left[\frac{1.3}{2.4}\right]^2.k^4 + \left[\frac{1.3.5}{2.4.6}\right]^2.k^6 +..\right]$$

$k=\sin(a/2)$ $\varphi_0=a$	$\varphi_0=\pi/2$	**Up to the 8th power**	= 1.177036285
	$\varphi_0=\pi/3$		= 1.073107004
	$\varphi_0=\pi/4$		= 1.039968689
	$\varphi_0=\pi/6$		= 1.017408711

The period T of the mathematical pendulum
A case of non linear 2nd order differential equation

s = distance = arc length

| s = L.φ | φ = f(t) | t = time |

From
$$0 < \varphi < \varphi_0 = a$$
to $0 < \theta < \pi/2$

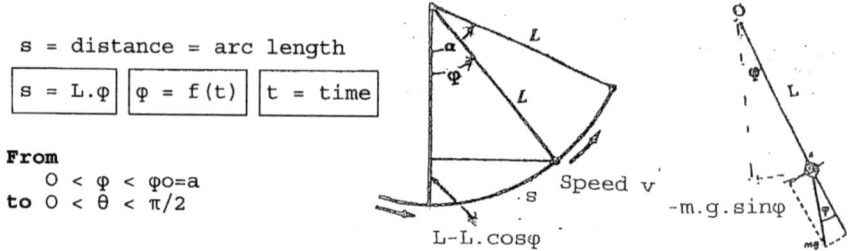

Speed v

L-L.cosφ

-m.g.sinφ

The pendulum consists of a bob at the end of a uniform string of length L. When displaced from the vertical by an angle a (a < π/2) and released from rest will oscillate with period T.

$$T = T(a)$$

The differential equation of the motion of a point mass (at the bob) according to **Newton's law** has the form:

| $m.L.[d^2\varphi/dt^2] = -m.g.\sin\varphi$ $0 < \varphi < \varphi_0$ | $\varphi_0 = a$ (vs) |

Or $\quad f''(t) + \dfrac{g}{L}.\sin[f(t)] = 0 \qquad$ | $\varphi = f(t)$ |

The function f(t) represents the oscillations of a material mass **m** (at the bob) being acted by an elastic force , whose magnitude is proportional to the deviation **φ** of the bob from the equilibrium position.

By the law of preservation of energy

Kinetic energy + Potential energy = C = Constant

- -

$$\frac{1}{2}.m.v^2 + m.L.g.[1-\cos\varphi] = m.L.g.[1-\cos a] \quad \boxed{0 < \varphi < \varphi_0 = a} \; 0 < \theta < \pi/2$$

- -

Therefore: $\quad v = \dfrac{ds}{dt} = L.\dfrac{d\varphi}{dt} = \sqrt{2.L.g.[\cos\varphi - \cos a]}$

- -

ie: $\quad dt = \sqrt{\dfrac{L}{2.g}} \cdot \dfrac{d\varphi}{\sqrt{\cos\varphi - \cos a}} \qquad$ This relationship leads to an elliptic integral

Change of the variable in the mathematical pendulum

From $0 < \varphi < \varphi o = a$ **to** $0 < \theta < \pi/2$

$$\sqrt{1-\omega^2} \quad \varphi/2 \quad 1 \quad \omega$$

$$\omega = k.\sin\theta$$

$$k = \sin(a/2)$$
$$k = \sin(a/2).\sin(\pi/2)$$

$$\sin\left(\frac{\varphi}{2}\right) = \omega \quad => \quad \cos\left(\frac{\varphi}{2}\right).d\left(\frac{\varphi}{2}\right) = d\omega$$

$$\int_0^{a/2} d\frac{\varphi}{2} = \int_0^k \frac{d\omega}{\sqrt{1-\omega^2}} = \int_0^{\pi/2} \frac{k.\cos\theta}{\sqrt{1 - k^2.\sin^2\theta}}.d\theta$$

$$\sqrt{\cos\varphi - \cos a} = \sqrt{\left[1 - 2.\sin^2\left(\frac{\varphi}{2}\right)\right] - \left[1 - 2.\sin^2\left(\frac{a}{2}\right)\right]}$$

$$\sqrt{\cos\varphi - \cos a} = \sqrt{\left[1 - 2.\omega^2\right] - \left[1 - 2.k^2\right]} \qquad \omega = k.\sin\theta$$

$$\sqrt{\cos\varphi - \cos a} = \sqrt{2.k.\cos\theta} \qquad\qquad k = \sin(a/2)$$

In conclusion:

$$\int_0^{a=\varphi o} \frac{d\varphi}{\sqrt{\cos\varphi - \cos a}} = \int_{\varphi=0}^{\varphi=a} \frac{2.d(\varphi/2)}{\sqrt{2.k.\cos\theta}} = \int_0^{\pi/2} \frac{2.k.\cos\theta}{\sqrt{2.k.\cos\theta}.\sqrt{1 - k^2.\sin^2\theta}}.d\theta$$

$$= \sqrt{2} . \int_0^{\pi/2} \frac{d\theta}{\sqrt{1 - k^2.\sin^2\theta}} \qquad \omega = k.\sin\theta$$

$$k = \sin(a/2)$$

And:

$$T = 4.\sqrt{\frac{L}{2.g}}.\left[\int_0^{a=\varphi o} \frac{d\varphi}{\sqrt{\cos\varphi - \cos a}}\right] = 4.\sqrt{\frac{L}{g}}.\left[\int_0^{\pi/2} \frac{d\theta}{\sqrt{1 - k^2.\sin^2\theta}}\right]$$

As an epilogue

Even though I am self educated in higher mathematics, by the meraki inspired to me by my uncle Fotis (a medical doctor) , I went deep into the various fields of higher mathematics.

Μεράκι means to love something very much,trying to perfect it. The desire and willpower to do so is very strong,overcoming a lot of effort , difficulties , and personal sacrifices.

The topics covered in my books were thoroughly understood , before written down in paper.

My ultimate goal is for somebody,who is really fond of mathematics,to grasp so many informations and details in a relatively short time period , something that took me many years.

Contrary to the common belief the topic of integral calculus is open to further improovement.

The John Bredakis method is along this direction

Also my observation that right angle triangles can in many cases bypass the inverse functions,is again in this direction

I dedicate this paper:

to my wife Sofia and my children Maria and Costas and to the sacred memory of my parents and uncle Fotis.

J.K.Bredakis MD

References:

1. **Higher Mathematics for beginners:**
   ----
   by Ya.B.Zeldovich
   (Mir Publishers Moscow 1973)

2. **Calculus with analytic geometry:**
   ----
   by Harley Flanders and Justin J Price
   (Academic Press 1978)

3. **A brief course of higher mathematics:**
   ----
   by V.A.Kudryavtsev and B.P.Demidovich
   (Mir Publisher's Moscow 1980)

4. **Concice Encyclopedia of Mathematics:**
   ----
   by W.Gellert,H.Kustner,M Hellwich,H Kastner
   (Van Nostrand Reinhold Company New York and other cities 1977)

5. **Computational Mathematics:**
   ----
   by B.P Demitovich and I.A.Maron
   (Mir publishers Moscow 1976)

6. **Advanced calculus:**
   ----
   by Leopold Flatto
   (The Wiiliams and Wilkins Company - Baltimore 1982)

7. **Mathematics Handbook for Science and Engineering:**
   ----
   by: Royal Lennart Rade and Bertil Westegren
   Fifth edition - 2004
   Springer Verlag Publications Inc
   Berlin - Heidelberg - New York

8. **Mathematical methods for physicists and engineers:**
   ----
   by: Royal Eugene Collins - 2nd corrected edition
   Dover Publications Inc - Mineola New York - USA 1991

9. **Differential Equations:**

   A systems approach - by: Jack Goldberg - and Merle C.Potter
   Prentice Hall International Editions
   Upper Saddle River , NJ - USA - 1998

- And a lot of personal work -